• THE R.C. RILEY COLOUR COLLECTION •

WESTERN REGION
DEVON & CORNWALL

IMAGES FROM THE TRANSPORT TREASURY ARCHIVE • COMPILED BY STUART MALTHOUSE

FRONT COVER • RCRC4399 Generations of holidaymakers have walked along the promenade at Teignmouth from the pier, past the fine tower of St Michael's church, and discovered the place where the main line through south Devon emerges from the cutting beyond the station and takes its place alongside the sea wall. Rail enthusiasts and non-enthusiasts alike have delighted in this spot, one of the most well-known and photographed locations on the UK rail system. Dick was there on 18 July 1958 and captured 7029 CLUN CASTLE on an unidentified up express but probably an inter-regional train for Liverpool or Manchester. CLUN CASTLE, named after one of the Welsh border castles in beautiful rural south Shropshire, was built by British Railways in May 1950 as one of the last batch of Castles, spending most of the 1950s working from Newton Abbot and Laira depots. It received a double chimney in October 1959, and was withdrawn in December 1965 from Old Oak Common, the last Castle to remain in BR service. However, it was to enter preservation, and after initial special train operations it became the flagship of the Tyseley collection. Today, after overhaul, 7029 is active once more on main line steam duties. There are more photographs and information on this famous location later in the book.

FRONTISPIECE • RCRC7305 If there are different opinions concerning the extent of the 'West Country' or the 'West of England', there can be little doubt about the meaning of Devon and Cornwall, two counties whose borders have been among the most stable throughout the last half century or more of administrative change. The Western's main line to the two counties enters Devon from Somerset in the depths of Whiteball Tunnel through a western outlier of the Black Down Hills. At the south end, the summit of the climb from Wellington is reached at Whiteball Siding Signal Box and the long descent to Exeter begins. This is the highest point on the original Bristol & Exeter Railway, opened through here in May 1844. In July 1956, 83D Laira-allocated No 4086 BUILTH CASTLE heads past the box with the 1325 (SO) Paddington–Kingswear. The train is made up of a typical rake of stock that would spend most of the time outside the summer months parked in sidings. Such trains offered a vital public service at the time but came to be an anathema to Dr Beeching. Whiteball Siding was not a public goods siding but an operational refuge siding for goods trains and banking engines which dropped off their trains here and awaited a return to Wellington.

OPPOSITE • RCRC4537 The most important station between Whiteball and Exeter until it was replaced by Tiverton Parkway in 1986 was Tiverton Junction. The immediate populated surroundings consisted of only the village of Willand and the station's principal function was as the interchange point for the two branches, to Tiverton itself, and the Culm Valley line to Hemyock. On 15 June 1962, 2-8-0 4700 class No 4705 attracts a group of onlookers as it prepares for departure with what is recorded as an up local. This loco was built in 1922 and withdrawn in December 1963, being one of nine locos designed by Churchward as an enlarged 4300 2-6-0 principally for overnight fitted goods work. They were also regularly used on this west of England main line on summer Saturday passenger trains. However, the date of this photograph is a summer Friday, and whether the appearance of this class of loco on an up local train was an exceptional circumstance or a regular return movement after working a down goods service is not recorded. Tiverton Junction was rebuilt with new island platforms on loops from the through lines in 1932. On the far left, the line behind the platform was used by the Tiverton branch train, while the corresponding one behind the down island on the right was for the Culm Valley trains. That branch is clearly visible climbing away into the trees in the centre of the photograph. The gated sidings on the left were added in 1943 for Air Ministry use.

ISBN: 978-1-913251-84-0
First published in 2024 by Transport Treasury Publishing Ltd., 16 Highworth Close, High Wycombe HP13 7PJ.

Printed by Short Run Press, Bittern Road, Sowton Industrial Estate, Exeter EX2 7LW.

INTRODUCTION

Railways, particularly steam, and photography were very much part of Riley family life. Richard had this enormous enthusiasm and expertise which resulted in a huge archive of colour slides, which have continued to be published since his death in 2006. His passion for this subject was never in doubt, but he was always modest about his place in being one of the Country's most talented and prolific railway photographers.

Often described as a gentleman, Richard was also a gentle man which often shone through when he was approached by other aspiring photographers and authors. He willingly gave his time to help others in many different fields. Equally, his family was very important to him. If ever advice and support were needed by the immediate family, his wisdom and integrity could always be relied upon, and he is much missed by us all.

Christine Riley

ABOVE • RCRC4647 On 23 July 1958, 1400 class 0-4-2T No 1449 is seen in the back platform of the down island at Tiverton Junction after bringing in a Culm Valley branch mixed train from Hemyock. The single passenger coach is joined by a rake of milk tanks. The Culm Valley Light Railway Company was authorised in 1873 as one of the few early light railways permitted by the 1868 act. It was opened in 1876 and worked by the Great Western which absorbed the company in 1880. The line was built as cheaply as possible to develop the agricultural and other economic activities in the upper part of the valley, although the passenger traffic was never expected to be substantial. After the establishment of a dairy products factory at Hemyock in the 1880s, the butter, cheese and other items from here became of increasing importance, and with the gradual establishment of liquid milk for the London market after the First World War, the output from the Hemyock factory became the dominant traffic on the line. Accordingly, although the passenger service on the branch ended in September 1963 and the general goods service two years later, the milk business kept the branch alive until final closure in 1975.

BELOW • RCRC3422 Dick did not always require locomotives and trains to be the main feature in his photographs. He also recorded other detail such as stations and general railway scenes. This study of the basic facilities at Coldharbour Halt on the Culm Valley branch was taken on 15 June 1962, just over a year before its closure. The name of the halt in the last sentence is not a misprint. The name used in most documents and timetables had one word which corresponded with the name of this area at the western edge of Uffculme as shown on Ordnance Survey maps. The two words on the station nameboard was the exception, and was an example of how many station names would appear differently in various contexts. The halt was opened on 23 February 1929 and a public goods siding was also provided which can be seen behind the halt. One of the most important customers for this facility would have been the owner of the large building behind the halt. This was Coldharbour Mill opened as a woollen textile mill in 1799 on the site of an earlier corn mill and one of the oldest such concerns in the world. Commercial production came to an end in 1981 and the mill is now a working museum.

LEFT • RCRC4658 This is Uffculme on the Culm Valley branch, also taken on 15 June 1962. The locomotive is No 1451, one of the Exeter-allocated Collett 1400 0-4-2 tanks. These were the dominant form of motive power on the line from the 1930s, being ideal for the light axle loading and curvature of the branch. This mixed train has no milk tanks on this occasion with grain appearing to be the most important commodity. The passenger carriage is one of the two ex-Barry Railway suburban brakes brought to the line in 1950 to replace even older vehicles. These were the last gas-lit coaches on British Railways, being converted from electricity, as the slow speed of the branch trains was not sufficient to recharge the electric batteries.

BELOW • RCRC4705 Another photograph from 15 June 1962 shows the branch terminus at Hemyock beside the River Culm and 1451 awaiting departure for Tiverton Junction with one of the Barry coaches. The dairy, the main reason for the line's continuing existence for many years, dominates the background. The siding on the left passed behind the station building, crossed the road beyond the station on a level crossing, and entered the dairy premises. Today, not only the railway but also the dairy is no more, the site of the latter being occupied by an estate of new houses.

PETER W GRAY, R C RILEY COLLECTION • RCRC4662 On 3 November 1962, an unidentified 1400 tank has arrived at the branch terminus of Hemyock with the 1342 departure from Tiverton Junction. By this date, for the last year of the passenger service, the two Barry coaches had been replaced by LNER suburban brakes, one of which forms the train seen here. These had to travel to Exeter at least weekly for the electric batteries to be recharged. As was customary, the few sidings at the little terminus were full of milk tanks, the one seen on the right standing on the line into the dairy. The standard six-wheeled glass-lined milk tanks had been introduced onto the branch from the early 1930s, to accommodate which some track improvements had been undertaken, indicating the prime importance of this traffic to the future of the line.

RCRC4310 Further towards Exeter on the main line is Stoke Canon, seen here looking south on 6 July 1957 as No 6829 BURMINGTON GRANGE (83A Newton Abbot) passes on the 1035 (SO) Paignton–Wolverhampton. This was another typical summer Saturday train but now in the year following which a decline in the enormous numbers of passengers travelling on such holiday trains to and from the west began. Indeed, it was only three weeks later on the 27th when what is generally recognised as the busiest (and most chaotic) day ever on the railways of Devon and Cornwall occurred. This was the second passenger station at Stoke Canon, replacing the first to the north (and nearer the village) in 1894. This change was made in order for the station to be served by the Exe Valley trains, the junction being to the south of the original site and immediately to the north of this station. The loops and new platforms, including a line for the Exe Valley trains on the far side of the up island, were installed in 1931, but the station was never a traditional branch junction and interchange point as the Exe Valley trains always ran through to Exeter.

PETER W GRAY, R C RILEY COLLECTION • RCRC4676 The Exe Valley line was opened from Morebath Junction on the Taunton–Barnstaple line to Tiverton in August 1884 where it joined the original 1848 branch from Tiverton Junction. The extension southwards to Stoke Canon followed in May 1885. At Tiverton, naturally the most important intermediate location, a new passenger station was provided when the Exe Valley line opened, the original branch terminus being located at the goods yard site. This station is shown here on 18 August 1963 with a 'full house' of trains, all powered by Exeter-allocated 1400 0-4-2 tanks, the staple motive power of the Exe Valley line and Tiverton Junction branch. On the left is 1471 in the bay platform with the 0848 to Tiverton Junction. In the through platforms are Exe Valley trains with 1451 on the left propelling the 0810 Dulverton–Exeter, while 1462 heads the 0755 Exeter–Dulverton. The Exe Valley services were withdrawn in October 1963 shortly after the date of this photo, but the Tiverton Junction service lasted until October 1964.

ABOVE • PETER W GRAY, R C RILEY COLLECTION • RCRC4624 The branch from Tiverton Junction to Tiverton was opened in 1848 but the single intermediate stopping place of Halberton Halt was not provided until 5 December 1927. The halt is seen here on 2 February 1963 during what was the worst Devon winter, at least in the twentieth century. 1421 is propelling the auto train away from the halt towards Tiverton. The bridge over the halt framing the photograph carried the lane which gave access from the village, but the overbridge to the west, part of which can be seen to the right of the locomotive, was far more interesting in that it was an aqueduct taking the Grand Western Canal over the railway.

OPPOSITE PAGE:

TOP • RCRC4681 Bampton was the last station going north on the Exe Valley line before Morebath Junction. The station is seen here looking south on 3 July 1963 from the overbridge which was centrally placed within the village served, a feature not found at so many Devon and Cornwall stations. 1400 0-4-2 tank No 1471 is about to leave for Dulverton with a service from Exeter. Along the face of the hillside in the background were extensive limestone quarries, the product being used for building and roadstone. At the time of this photo the quarries were still active, as evidenced by the smoke or steam rising above the goods shed. Until the 1950s much of the stone was transported by the railway. A siding connected the goods yard to the westernmost quarry and the overgrown remains of this can be seen immediately to the left of the goods shed. In earlier years a horse and gravity worked narrow gauge tramway brought stone from the other quarries to the east down to the standard gauge siding.

BOTTOM • RCRC3493 Although the Exe Valley line joined the Taunton to Barnstaple line physically at Morebath Junction, the passenger trains continued a further two miles towards Barnstaple where interchange between the services on the two lines was made at Dulverton. Because of this, arguably Dulverton was the most important station between Taunton and Barnstaple although some of the others served a larger surrounding population. Half a mile before the station the line crossed the River Exe from Devon into Somerset, but Dulverton can justifiably be included here as the effective terminus of the Exe Valley line. The station is seen here on 15 June 1963 with another 'full house'. On the left beyond the interchange running-in board is the rear of the 1610 Barnstaple Junction to Taunton train hauled by an unseen No 6372. To its right in the down island is No 7304 on the 1620 Taunton to Barnstaple Junction. The station is a scheduled passing place on the single-track route. Both locos are Taunton-based 4300 class 2-6-0s, the staple motive power of the line's services. On the right of the down island, 1400 class 0-4-2T No 1421, having made connections, will propel the 1715 to Exeter via the Exe Valley line.

PTON
1471

ABOVE • RCRC4169 Taunton-allocated Modified Hall No 6995 BENTHALL HALL passes the Cowley Bridge Inn at Cowley Bridge Junction on 5 July 1961 with a down stopping service. The bracket signal beyond the train has quickly been returned to danger. The subsidiary arms are for the diverging goods lines and down loop which begin beyond the road overbridge off the photo to the right. Curving away from the junction in the foreground over separate courses of the River Exe is the Southern line to its empire west of Exeter. This area, where the River Creedy joins the Exe, was notorious for flooding during heavy Devon rains until improvements in more recent times. Cowley Bridge can probably be considered the most notable junction in Devon and Cornwall throughout railway history. It began life in 1851 with the opening of the broad gauge Exeter & Crediton Railway branch to Crediton worked by the Bristol & Exeter Railway. In the normal course of nature, as happened to hundreds of other such lines throughout the UK, the line would have passed to the B&ER and hence to the Great Western. However, although it was no nearer than Salisbury and Dorchester at the time, the London & South Western Railway saw the E&CR and the further proposals to extend to Barnstaple, as pawns to achieve its share of colonising Devon and Cornwall. Accordingly, the E&CR became involved in the Gauge War and, the standard gauge being victorious in this case, was eventually leased to the LSWR in 1862 and absorbed in 1879. The railway history and geography of Devon and Cornwall was thus transformed, and the original E&CR branch became the access to all the LSWR and later Southern lines beyond Exeter. However, with the failure of the subsequent schemes to avoid Exeter St Davids with a line of its own joining the route off the photo to the left, the Southern owners were always left with the handicap of using the busy GWR main line for one and a half miles from the south end of Exeter St Davids station to Cowley Bridge Junction in order to reach its system further west, the so-called 'Withered Arm'.

RIGHT • RCRC4596 The road overbridge at Cowley Bridge Junction is seen on 16 July 1958 as 2884 class 2-8-0 No 3864 (83A Newton Abbot) crosses the junction and comes through with a down mixed goods. The line in the foreground and that just seen diverging on the far right is the start of the separate tracks leading to Riverside Yard and the goods lines avoiding the platforms at Exeter St Davids. The start of the down loop is off the photo to the right. The junction signal box is out of sight behind the road bridge on the up (left) side.

BELOW • RCRC4665 In the 1950s, Exeter St Davids would probably have been voted in the top ten of 'Resorts for Railfans', especially on summer Saturdays when the volume and variety of long distance passenger services was unsurpassed. The north end is seen here on 5 July 1961 with 1400 class 0-4-2T No 1451 awaiting departure from the bay platform 2 with an Exe Valley train. The opposite platform, No 1, handled all down Western Region trains which called at the station, although there was a through line beyond this where non-stops could overtake. Beyond the platform is the Red Cow Level Crossing which gave road access to the goods yards and loco depot on the west side of the station and through to Exwick on the far side of the River Exe. Exeter Middle Signal Box is predominant and further on the smaller of the two goods sheds. The end of the larger shed alongside the station can be seen through the windows of 1451. Out of sight behind the loco and off the photo to the left are the other passenger platforms, the central island numbers 3 and 4 used by Southern trains and the western island platforms 5 and 6 for Western Region up services. On the far side of the level crossing can be seen the banking engine's siding where locos waited their turn to assist Southern trains from platform 3 up the 1 in 37 to Exeter Central. By this time class Z 0-8-0Ts had replaced the E1/R 0-6-2Ts and M7 0-4-4Ts on this work, one of which can be seen awaiting its next duty.

RCRC4344 Castle class 4-6-0 No 5096 BRIDGWATER CASTLE (82A Bristol Bath Road) leaves platform 1 at Exeter St Davids on 19 July 1958 with the Liverpool to Plymouth through train. The main station buildings on platform 1 are notable for the row of round-headed windows to the offices on the first floor. The loco is about to cross the tracks used by Southern trains, leading from the central island platforms 3 and 4 on the left to the incline up to Exeter Central off the photo to the right. The building on the right is a carriage shed which extends alongside the incline. The through line between platforms 1 and 3 used by non-stopping down Western Region services can be seen immediately to the left of the train.

ABOVE • RCRC7272 On 23 June 1962, Castle class No 4037 THE SOUTH WALES BORDERERS is seen at the southern entrance to Exeter loco shed alongside St Davids station. The running-in board is that on the up western island platforms 5 and 6, and behind can be seen the back of the carriage shed alongside the Southern incline. 4037 had been allocated to Newton Abbot but after the closure of 83A to steam, was actually at Exeter by this date. It was built in December 1910 as Star class QUEEN PHILIPPA, and rebuilt as a Castle in June 1926. After a special request, it was renamed for the regiment in April 1937 at a ceremony at Paddington. The loco was withdrawn shortly after the date of this photo in September 1962 having run nearly two and a half million miles, one of the highest totals of any steam locomotive, and the record for a Great Western one. It was scrapped at Cashmore's at Newport in the December.

OPPOSITE • RCRC0596 A view over Exeter loco depot and yard, taken from the water tower on 2 July 1963. On the left are two lines of locomotives awaiting the final call to the scrapyard. Among them are Southern engines including an M7, a couple of 700 class 0-6-0 'Black Motors' and an N class. Prior to the regional boundary change on 1 January 1963 these locomotives would have been dumped at Exmouth Junction. A variety of Western classes occupy the roofless shed and yard, but time was running out as it would close to steam that October. The site continued as a diesel stabling point. The new order in the shape of a Hymek diesel hydraulic with a mixed freight passes on the goods lines avoiding the passenger platforms off to the right, and part of the goods yard can be seen beyond it. Had the Southern's Exeter St Davids avoiding line materialised it would have crossed the Western lines on a viaduct roughly in this position and curved round to follow the fields beyond the shed northward to Cowley Bridge in the dip between the hills on the horizon. Presumably interchange platforms would have been provided behind the shed but it would have been a long walk, most likely using a lengthy footbridge from the main station in Gloucester Central to Eastgate fashion.

RCRC4303 Looking in the opposite direction from the water tower over the southern approach to Exeter St Davids on 23 June 1962. Approaching on the St Davids Bridge over the River Exe and signalled into platform 6 is Modified Hall No 7909 HEVENINGHAM HALL with an up express. Meanwhile, 6827 LLANFRECHFA GRANGE passes on a down van and parcels train. On the left is Exeter West Signal Box, the largest and busiest of the Exeter boxes, controlling not only the passenger lines through the south of the station, but the Southern incline, the junction of the goods avoiding lines, and the entry to the loco depot. A new frame with 131 levers was installed in 1958. On 1950s summer Saturdays the box was manned by two signalmen and a booking clerk. The box was closed in 1985 with the West of England resignalling but fortunately was preserved by the Exeter West Group. After an heroic effort of dismantling, transport, and reassembly it is now a working exhibit at Crewe Heritage Centre. Behind the box the 1 in 37 of the Southern incline to St Davids Tunnel and Exeter Central is clearly evident. Had the avoiding line come to be, this scene would have been transformed. It would have left the incline above the signal and crossed the layout in the foreground on a viaduct.

LEFT TOP • RCRC3622 This is an undated view; Taunton-allocated 4300 class 2-6-0 No 6377 is most likely on a summer Saturday service as a change from its normal duties on the Barnstaple line, passing the signal box and platforms at Dawlish Warren. This station is the second to serve the small resort at the mouth of the Exe estuary, being opened in 1912 to replace a smaller halt from 1905, a quarter mile to the south.

LEFT BOTTOM • RCRC4408 No 7031 CROMWELL'S CASTLE takes the curve from Dawlish Warren and approaches Langstone Rock on 3 July 1957. Among the last batch of Castles built by British Railways in June 1950, the loco was allocated to Laira (83D) at the time. It was named after an artillery fort well beyond the railway system on Tresco in the Isles of Scilly, first constructed in 1651 at the end of the Civil War to protect the islands after they had been occupied by Parliamentary forces. The train is the 0530 Paddington–Penzance via Bristol. This was a service which called at, and was essentially for, many intermediate stations, including all of those in Cornwall; arriving in Penzance at 1625, only half an hour before the Cornish Riviera Express which had left Paddington five hours after it. On the right is the start of the sea wall path to Dawlish and the breakwater protecting the warren, which stretches more than half way across the Exe estuary.

OPPOSITE • RCRC7387 The South Devon Railway reached the sea after Dawlish Warren by cutting through the headland of Langstone Cliff. This left the 'island' of Langstone Rock, from the heights of which a grandstand view of the line could be enjoyed and which enabled photographs such as this. The subject here, on 3 July 1957, is the up 'Torbay Express', the 1125 from Kingswear to Paddington, which only called at Exeter St Davids after Torquay. The loco is Newton Abbot-based Castle class No 5059 EARL ST ALDWYN which was built in May 1937 as POWIS CASTLE and renamed in the October. The decision to use chocolate and cream coaches on the Western Region's named trains in 1956 was a splendid one, but all too often the appearance of a coach or coaches in a different livery following maintenance requirements or other circumstances, as here, spoiled the effect. One of the most famous sea-side walks in the country, for all, but especially for railway enthusiasts, leads in the distance to Dawlish station and town in the haze.

520

BEACH
HOTEL

OPPOSITE PAGE:

TOP • PETER W GRAY, R C RILEY COLLECTION • RCRC7427 Love it or loathe it, no one can escape the railway at Dawlish. It assumes the place of a road promenade with only the sea-wall path between it and the beach. On Marine Parade on 27 August 1961 (a Sunday), strollers display the usual mixture of interest and indifference as Castle No 5087 TINTERN ABBEY runs out of Kennaway Tunnel towards the low Colonnade Viaduct over the access to the beach and the station stop with the 1710 Kingswear to Bristol. The loco, based at 81A Old Oak Common, was built in November 1940 but used the frames of Star class No 4067 TINTERN ABBEY dating from January 1923, and so strictly can be classified as a rebuild.

BOTTOM • W POTTER, R C RILEY COLLECTION • RCRC4125 Between the Dawlish and Teignmouth stretches of sea wall, the main line cuts through a stretch of headlands, cliffs, and coves with the help of five tunnels — Kennaway, Coryton, Phillot, Clerks, and Parsons. It is an area where each part is only reached by footpath, and on 24 June 1961 Dick ventured above the longest open stretch of line between the Clerks and Parsons Tunnels. Emerging from the former is No 6926 HOLKHAM HALL with a train for Paignton. Beyond the train between the breakwater and Horse Rocks is the small beach of Shell Cove, beyond which is Horse Cove and Coryton's Cove with Dawlish and its station visible in the background.

ABOVE • RCRC3711 The Teignmouth sea wall begins at the south end of Parsons Tunnel by Holcombe Beach. As a change from the diet of long distance trains, this is a view of No 4117, a 5101 class Large Prairie 2-6-2T (83C Exeter) emerging from the tunnel on 14 July 1959 with an Exeter to Newton Abbot local service. On the left is Parsons Tunnel Signal Box. In the 1890s the box here controlled the end of the double line section from Teignmouth when the route was only single line through the tunnels. After the 1905 doubling through the tunnels (the last section of the main line to be so improved), the box closed, but a second box, the one shown here and a little further from the tunnel, was opened in 1934 solely to break up the section between Dawlish and Teignmouth. This was most beneficial on summer Saturdays when there was a constant procession of trains, sometimes in a stationary queue if there was congestion at Newton Abbot. The box finally closed in 1964.

ABOVE • RCRC7395 Dick's photographs on the Western lines in Devon and Cornwall ranged widely including many lesser-recorded locations. However, he loved the familiar surroundings of the Teignmouth sea wall as much as any other photographer and captured many images here. Approaching Parsons Tunnel on 14 July 1959 is the up 'Torbay Express', this time with a uniform rake of chocolate and cream coaching stock. The loco is Old Oak Common's No 5065 NEWPORT CASTLE. Holidaymakers enjoy the beach during one of the hottest of twentieth century summers, and at the top of the steps a youngster admires the passing spectacle. Did it engender a permanent enthusiasm?

OPPOSITE PAGE:

TOP • RCRC4255 Mid-way between Parsons Tunnel and the curve away from the sea towards Teignmouth station is Sprey Point, a small peninsular extension of the wall complete with grass and bushes. In this view from 1 July 1957 looking back towards Parsons Tunnel, Taunton-allocated Collett Hall No 4971 STANWAY HALL approaches the point with an Exeter to Paignton stopping train.

BOTTOM • RCRC4251 On 14 July 1959, another Taunton-allocated Hall, No 4970 SKETTY HALL, passes Sprey Point with a local passenger train. The milepost indication of 208 and a quarter is from Paddington via Bristol on the original Great Western, Bristol & Exeter, and South Devon Railways main lines.

4970

ABOVE • RCRC7327 Looking in the opposite direction towards Teignmouth, No 5011 TINTAGEL CASTLE (83A) is seen alongside Sprey Point with the 0730 (SX) Penzance to Manchester through train. At this time such services were routed via the Severn Tunnel and the Hereford–Shrewsbury, North and West route. 1 July 1957.

BELOW • RCRC4081 On the same day, royal progress is made past the Parsons Tunnel up distant signal at Sprey Point by the last of the Kings, Laira's No 6029 KING EDWARD VIII. Built in August 1930 and originally named KING STEPHEN, the loco was renamed in May 1936 for the new monarch, although with the latter's abdication, 6028 had to go through a similar procedure less than a year later. The loco would be fitted with a double chimney in December of this year. The train is the 0625 Penzance to Paddington made up of a variety of stock, the first of which is a Collett E150 centenary stock gangwayed brake composite, one of six built in July 1935. Teignmouth, with its pier and church, can be seen beyond where the line curves inland towards the station.

LEFT • RCRC4237 Nearer Teignmouth on 14 July 1959, 4900 Hall class No 4949 PACKWOOD HALL (83B Taunton) passes with an up goods. The down line here has been relaid with flat bottom rail, whereas the older bull-head rail is still in use on the up line. On the left by the wall, the Teignmouth up starter signal is located on the down side of the line to assist with sighting from trains coming round the curve from the station.

BELOW • RCRC3640 As mentioned in the caption for the cover photograph, the curve from the cutting leading from Teignmouth station on to the sea wall must be one of the most photographed locations on the whole of the UK rail system. On 1 July 1957, 4300 class No 6394 comes round the curve with an up local train. When this section of the South Devon Railway was first opened in May 1846, the line was single track and curved here into East Cliff Tunnel. Between 1882 and 1884 the tunnel was opened out to form the cutting and the line was doubled eastwards to Parsons Tunnel. The fine girder bridge, the most characteristic feature at this location, was built to carry Eastcliff Walk, which formerly had passed over the tunnel.

RCRC7384 On 2 July 1957, 5059 EARL ST ALDWYN (83A) comes onto the sea wall at Teignmouth with 'The Devonian' from Paignton to Bradford Forster Square. This was the one service from the Derby–Birmingham–Bristol main line in the 1950s which on a daily basis outside of summer Saturdays crossed the rigid regional divide at Bristol, from the London Midland to the Western regions. To the present author, a ten year old Rotherham trainspotter towards the end of the decade, 'The Devonian' meant Bristol Barrow Road and Leeds Holbeck Jubilees. That the same train could daily pass such exotic locations with such exotic motive power was not yet something which could be personally experienced, but at least we had a red headboard!

RCRC7366 Taken from the road bridge across the River Teign estuary to Shaldon, Laira-allocated Castle class No 5049 EARL OF PLYMOUTH leaves Teignmouth on Friday 5 July 1957. The train is the working of restaurant cars complete with provisions and staff which left Paddington at mid-day on Fridays to be added to up summer Saturday workings from Plymouth and Cornwall the following day. This must be the ultimate gesture of public service, taking priority over economic working which within a few years would be history. Today such an activity seems to belong to the standards of a different civilisation! Perhaps better known as a holiday resort, Teignmouth has always been a significant port and part of the rail-served quays can be seen on the right. In the 1950s and 1960s, exports reached some 400,000 tons of which three quarters was ball clay from the Bovey basin with agricultural products making up most of the remainder. Imports included fish, coal, and timber, while ship and boat building was undertaken.

ABOVE • RCRC4160 In this photo from 17 July 1959, Hawksworth Modified Hall No 6988 SWITHLAND HALL (83D Laira) is in charge of a down mixed goods. The location is almost at the head of the Teign estuary a mile and a half before Newton Abbot.

BELOW • RCRC7281 On 15 July 1959, the up 'Torbay Express' passes beneath the bridge carrying Hackney Lane from Kingsteignton down to the Passage House Inn by the bank of the River Teign just before it reaches the head of the estuary. The loco is Newton Abbot's No 4098 KIDWELLY CASTLE. The chocolate lettering and cream background locomotive headboards and carriage roofboards were introduced to match the new coach liveries in 1956. The headboard for the 'Torbay Express' had the arms and motto of Devon County Council between the two words of the title, the motto being 'Auxilio Divino' (by Divine Aid).

332

RCRC4427 On the same day and at the same location as photo 4160 on the previous page, an unidentified Castle is seen with the down 'Torbay Express', again consisting of a complete set of chocolate and cream coaches.

RCRC4256 On 19 July 1958, No 4980 WROTTESLEY HALL of 82B Bristol St Philips Marsh departs from Newton Abbot and approaches the massive East Signal Box, crossing over the junction with the Moretonhampstead branch which passes behind the box. The train is the 0905 Kingswear to Swansea High Street summer Saturday service yet consists largely of ex-Southern and LNER coaches in both crimson and cream and maroon liveries.

5880

6026
6029

OPPOSITE • RCRC4108 Long-term Laira-allocated King, No 6026 KING JOHN is seen at Newton Abbot on 3 July 1957 with an unidentified up express. Taken from the Moretonhampstead bay platform, the train is seen in platforms 7/8 on the officially termed up main line on the western side of the up island platform. The particular lighting conditions make the green locomotive appear almost in mixed traffic black livery. KING JOHN would receive a double chimney in March 1958. No 6829 BURMINGTON GRANGE can be seen on the far side of the island, the up relief line in platforms 5/6. A Large Prairie is on the far side of the down island, the down relief line, most likely on banking duties at the rear of a train in platforms 1/2. Beyond can be seen the main building of Newton Abbot Works. The rebuilding of Newton Abbot station, originally planned before the First World War, was finally completed in 1927. This included the extended island platforms and the wide footbridge with both stairs and lifts seen above the canopy.

ABOVE • RCRC4110 On 14 July 1955, super power occupies platforms 3/4, the down main, at Newton Abbot with a Paddington to Penzance express. The locos are King No 6025 KING HENRY III (Laira 83D) and Castle No 5059 EARL ST ALDWYN (83A Newton Abbot). The Castle, being allocated to the adjacent MPD, would have just been put on to the train as pilot to assist over the south Devon banks. In continuing GW tradition, this is 'inside' the train engine, a practice requiring extra moves each time such pilots are attached or detached. Hall No 4949 PACKWOOD HALL (83B Taunton) occupies the down relief in in platforms 1/2 most likely with a connecting Torquay line train. Further locomotives are visible in the shed yard of 83A to the right, beyond a row of coal wagons.

RCRC4523 In the summer of 1961, on Saturday 15 July, Newton Abbot is seen in transition with the dieselisation of the Western Region in Devon and Cornwall well under way. The old order is seen in platforms 1/2 with 4700 class 2-8-0 No 4704 (81A) on the 1320 (SO) Paddington–Kingswear. This train, usually rostered for a 4700 at the time, acted as a relief 'Torbay Express' and at least on this occasion appears to be usurping the main train's rake of chocolate and cream coaches complete with carriage roofboards. On the right, progress is represented by a D6300 type 2 diesel hydraulic, a diesel shunter, and a diesel multiple unit.

RCRC7396 Six weeks later, on 26 August 1961, the 'Torbay Express' itself is seen passing Newton Abbot non-stop on the down through line. The locomotive is 81A Old Oak Common-allocated Castle No 5066 SIR FELIX POLE. This was built in July 1937 as WARDOUR CASTLE and renamed in April 1956 for the 1921 to 1929 Great Western Railway general manager, following his death in the January of 1956. It was fitted with a double chimney in April 1959 and withdrawn in September 1962. On the right are two of the D6300 diesel hydraulics, Nos D6319 and D6335. These locomotives by now were well established on local passenger and goods workings, including piloting work over the south Devon banks.

RCRC4436 On 18 July 1958, No 7809 CHILDREY MANOR is seen with a Torquay line local service between Newton Abbot station and Aller Junction. Like most Manors, 7809 had a variety of allocations but at this time was a Newton Abbot asset. The four coaches from the loco are: a Collett sunshine stock D121/124 gangwayed brake third of 1935; a Collett bow-ended gangwayed third; a Collett sunshine stock composite; and a Collett bow-ended brake third. The separation of the main Plymouth and Torbay routes at Aller Junction has been, at various times, a physical junction of the tracks or simply a geographical divergence of routes with connections made at Newton Abbot station, as it is today. At the time of the photo the connections at the junction dated from 1941 and the Manor is seen on the down relief line. In succession, the lines in front of the train, a little unusually, are down main, up relief, and up main, and this order was maintained through Newton Abbot station. The main lines were principally used by Plymouth trains and the relief lines by the Torquay and Kingswear services but the connections at the junction enabled flexibility. On the A380 road from Newton Abbot to Torquay a Devon General bus parallels the train.

RCRC4052 The climb to Dainton starts shortly after Aller Junction, and this view is on the 1 in 46 approaching Stoneycombe Quarry nearly half way up the almost three mile ascent. Brunel has been criticised for leaving a legacy of heavy gradients because they would have had a lesser effect on the proposed atmospheric traction than on steam power, but there is no doubt that whatever course had been taken across this country the gradients would not have been easy. The 1130 (SX) Paddington–Penzance express is hauled by County No 1011 COUNTY OF CHESTER piloting King No 6016 KING EDWARD V, with the pilot this time in the more conventional position. The King is an Old Oak loco, but the County is from 82A Bristol Bath Road and has perhaps been seconded after arrival at Newton Abbot for assisting work on the south Devon banks. 3 July 1957.

1011

THE CORNISHMAN
825

RCRC7418 On the same day at the same location, the down 'Cornishman' is seen behind Newton Abbot Castle No 5079 LYSANDER. Originally LYDFORD CASTLE, this was one of the class renamed after World War Two aircraft, in this case in November 1940. At this time, 'The Cornishman' was an internal Western Region responsibility, the down train being the 0900 Wolverhampton Low Level to Penzance via the Stratford upon Avon to Cheltenham line. Accordingly, it was one of the named trains chosen for a livery change in 1956. The crest on top of the headboard is a Cornish pixie vaulting over a toadstool.

RCRC4280 On Saturday 19 July 1958 the 1115 (SO) Newquay–Wolverhampton descends Dainton bank past Stoneycombe Quarry. The train is double-headed by County No 1016 COUNTY OF HANTS with Grange No 6808 BEENHAM GRANGE as pilot. The latter is one of Penzance's Granges but the County is a less frequent passer-by at this location as it belonged to 84G Shrewsbury at the time. Stoneycombe Quarry produced pink limestones of Devonian age. As well as being used in a wide variety of applications common to all aggregates, their nature meant they were also used for decorative purposes, giving a marble effect. The loop and sidings for rail access to the quarry are clearly visible in the photograph. For many years the quarry was also served by an internal two-foot gauge tramway.

ABOVE • RCRC7365 The down 'Royal Duchy' is seen assaulting the final 1 in 36/44 of Dainton between Stoneycombe and the tunnel on 1 July 1957. Power is provided by Newton Abbot's No 5049 EARL OF PLYMOUTH, although with no noticeable exhaust in the warm air. 'The Royal Duchy' was one of the Western Region's new titled trains introduced in the mid-1950s, possibly as an excuse to employ more chocolate and cream coaches. The inaugural service was on 28 January 1957. However, it was not the fastest of trains with several intermediate calls. The 1330 down departure from Paddington arrived in Penzance at 2100, and the 1105 up working took some half an hour longer, both being over an hour slower than the corresponding 'Cornish Riviera Express'. The locomotive headboard incorporated the arms of the Duchy of Cornwall, to use which and the title generally required the permission of HM The Queen.

OPPOSITE • RCRC4090 The summit of the climb from near Aller Junction is reached on a short level section within Dainton Tunnel. Immediately at the western end of the tunnel, as seen here on Saturday 29 June 1957, the equally precipitous curving descent to Totnes in the Dart Valley begins. The sidings either side of the main lines are on the level, so the severity of the gradient can easily be seen. The up siding on the left is a refuge for the bankers, whereas the down sidings passing behind Dainton Sidings Signal Box included a banker's siding as well as two serving an adjacent quarry. The train is the 1035 (SO) Paddington–Penzance double-headed by Old Oak Common Kings Nos 6013 KING HENRY VIII and 6028 KING GEORGE VI. Such mighty 16P power was not a freak one-off occurrence. On summer Saturdays during most of the 1950s, the 1030 'Cornish Riviera Express' made its first public call from Paddington at Truro, carrying only passengers for the far west, principally St Ives and Penzance. An operating stop was made at Newton Abbot to replace its King with two lighter 4-6-0s which could then work through to Cornwall without a further Plymouth stop, Kings being barred west of Keyham. The displaced King was then added as assistance over the south Devon banks to the following relief 1035 which called at Plymouth and served other Cornwall stations. As this train was also usually King-hauled, this double-headed spectacle was a regular summer Saturday sight.

RCRC4627 A group of youngsters admire the branch train for Totnes at the Ashburton terminus with its timber overall roof on 2 July 1957. The usual 1400 class power for the Hawksworth and Collett auto coaches is provided by No 1427 on this occasion. Closing to passengers in 1958 and goods in 1962, the branch became available for one of the earliest standard gauge preservation schemes. At first it was hoped to reopen the entire branch for passengers, but the section above Buckfastleigh was lost to the A38 trunk road improvement scheme and was only used by the Dart Valley Light Railway for stock storage and works trains for a limited period. The station, incorporating the roof, became the Station Garage which still exists today, but there remains alive a hope that a realigned railway can once more reach Ashburton and use the original terminus.

1427

PETER W GRAY, R C RILEY COLLECTION • RCRC6023 One result of choosing the southern fringes of Dartmoor for the course of the South Devon Railway main line was that the broad agricultural area of the South Hams southwards to the coast, which could have provided an alternative route, was left without easy rail access. The first successful scheme to serve the area, and not until 1893, was the long and sinuous branch from the main line at Brent to Kingsbridge. Indeed it was the only line to enter the South Hams, apart from that to Yealmpton opened five years later which had originally been proposed to go as far as Modbury. The substantial branch terminus at Kingsbridge is seen here on 8 June 1961. The goods yard is on the left with the goods shed off the photo, and in the centre is the unusual carriage shed. The passenger terminus had two platforms, although the shorter was principally for serving the loading dock as evidenced here. On the right can be seen part of the single road engine shed. In the main platform, the branch train is hauled by Newton Abbot's 4500 class Small Prairie No 4561 which is now preserved at the West Somerset Railway. The Kingsbridge branch closed for both passengers and goods in September 1963.

BELOW • RCRC7321 The summit of the main line is reached at Wrangaton and, following the boundary of the Dartmoor National Park, the small town of Ivybridge is reached on the relatively easy first stage of the descent to Plymouth. At a pleasant although not often photographed location, No 4096 HIGHCLERE CASTLE (82A Bristol Bath Road) is seen on 4 July 1957 with an up express. Highclere Castle, south of Newbury near the Didcot Newbury & Southampton line, is in fact a seventeenth century country house renovated in the 1840s for the Earls of Caernarvon. In recent years it has gained international fame as the location for the television series 'Downton Abbey'. On the far left beyond Ivybridge Viaduct over the River Erme can be seen Ivybridge passenger station which closed in 1959. The goods yard off the photo further left only opened in 1911, replacing one behind the photographer further to the east. For some years china clay brought down by road from Lee Moor was transferred to rail here. In the last few decades Ivybridge has grown as a Plymouth commuter town and in 1994 a new station was opened to serve this business, but further east beyond the original goods yard.

RCRC3514 The eastbound climb to Wrangaton begins in earnest at Plympton with the first three miles to Hemerdon Siding being the most severe at 1 in 41/47/42. On 5 July 1955 an unidentified up express approaches the Hemerdon Siding distant signal at milepost 239 and three quarters, some 50 chains before the siding and summit of this steepest section of the climb. Hemerdon Siding closed as a public goods siding as early as 1902 but survived as an operational location with passing loops for goods trains and as refuges for the bankers. Dick did not record the identity of the locos, but they are certainly a Modified Hall piloting a Castle. The Hall could be 7909 HEVENINGHAM HALL, a Laira locomotive at the time. The relaid down line contrasts with the old bull-head rail surviving on the up track.

ABOVE • RCRC4053 On 15 July 1958, Hawksworth County 4-6-0 No 1025 COUNTY OF RADNOR (84G Shrewsbury) passes through Plympton station at the start of the climb to Hemerdon, which it is tackling unassisted or possibly with the help of a banker. The train is the through daytime service from Penzance to Manchester, with the restaurant car and first coach, both of LMS origin, being added at Plymouth. Between 1904 and 1930 Plympton was the eastern terminus of the intensive Plymouth area suburban services, initially using steam railmotors and later auto trains. The station closed in 1959 but there is a proposal to reopen it as part of a Plymouth area metro-style service once more.

OPPOSITE PAGE:

TOP • RCRC3959 Laira-allocated 5700 class 0-6-0PT No 4679 crosses the River Plym on 29 August 1961 after passing through Tavistock Junction with a train of china clay from Marsh Mills, most likely destined for export from Fowey. The Tavistock Junction bracket signal for the up line carries arms for, from left to right, the Tavistock and Launceston branch, access to Tavistock Junction Yard, and the main line through to Plympton. The overbridge with ornate railings beyond the signals carries the private road to Saltram House, the former home of Lord Morley, owner of the Lee Moor Tramway, and now in the care of the National Trust. Between 1970 and 2001 the preserved Peckett 0-4-0ST Lee Moor No 2 which worked on the tramway was on display there. Regrettably, today this view is spoilt as the higher and larger flyover of the A38 Devon Expressway overshadows the bridge.

BOTTOM • RCRC3815 At the same location and on the same day, No 1363 crosses the River Plym with a (probably empty) working of mineral wagons for Tavistock Junction Yard or Marsh Mills. No 1363 was one of the five Churchward 1361 class 0-6-0 saddle tanks built in 1910 with short wheelbases for shunting in docks and similar locations, and replacing much older locomotives. There were generally at least three of the class based at Laira for work in Plymouth's Millbay Docks and Sutton Harbour, but they could also be found on local goods trip workings, as seen here. 1363 was the sole survivor at Plymouth by this date. It was withdrawn in December 1962 but entered preservation in the care of the Great Western Society at Didcot. The River Plym appears to be in full flood, presumably after typical summer Devon rains, and appears to be carrying a fair amount of milky china clay waste from Lee Moor. The tower on the hill in the background near Egg Buckland marks the present location of the Plymouth Snowsports Centre, in an area now completely developed with housing and an industrial and retail park.

RCRC3960 At 32 miles, the Tavistock Junction to Launceston branch was the longest of the GW branches from the main line west of Whiteball, penetrating deep into what later became more familiar as Southern territory. It was opened in two stages, to Tavistock in 1859 and on to Launceston six years later. The first station on the line was Marsh Mills, around the corner from Tavistock Junction. Activity here greatly increased from 1921 when new china clay dries were opened receiving clay slurry piped from the pits at Lee Moor. This facility led to a gradual decline of traffic on the original Lee Moor Tramway but was of benefit to the GWR, as most of the clay was railed for export through Fowey. In this view from 29 August 1961, 8750 class pannier tank No 4679 is seen in the station during the shunting and preparation of a china clay train. The dries are seen in the background with the line serving them ascending beyond the station from the sidings on the right. The Launceston branch was originally broad gauge and Marsh Mills is one of those stations where the platforms have never been widened or resited, so that after the conversion there remained a legacy of an extended '6 foot' between the tracks.

RCRC5916 Shaugh Bridge Platform, further up the line beyond Bickleigh, would surely be a candidate for one of the most delightfully located and characteristic Great Western halts. Sylvan surroundings and single platform with the familiar pagoda shelter – what could be better, even if by this time the halt was looking a little run down? It was added on 21 August 1907 to serve the hamlet of Shaugh Prior and for tourists to reach the picturesque section of the valley around the Dewerstone Rock. The line followed the Plym Valley up from Marsh Mills, and here was some way above the river down to the right. From here, the line then followed the valley of the tributary River Meavy, to reach which it passed through the slopes in the background by means of Shaugh Tunnel, whose southern portal can be glimpsed to the left of the prominent conifer tree. Slowing for the stop with a Launceston to Plymouth train on 29 September 1962, although apparently without any passengers to pick up, is 4500 class Small Prairie No 4555, the normal type of motive power for the branch service. This is another of today's preserved members of the class, carrying the name WARRIOR and based on the Dartmouth Steam Railway.

BELOW • RCRC3516 The Launceston line had its own branch from Yelverton, but this was no conventional rural line. Certainly the lower section passed through fields, woods, and by the picturesque Burrator Reservoir. But once across Peek Hill Bridge over the B3212, the line entered the wilderness of Dartmoor to twist and climb at 1 in 40 to its Princetown terminus, at 1373 feet the highest railway station in England. Opened in 1883 and using much of the trackbed of the early Plymouth & Dartmoor Railway, the line carried visitors and supplies for the prison as well as general goods for the small town, and served several small quarries on the way. As well as being used by the local inhabitants, in later years the passenger traffic was supplemented by walkers and other tourists. Ingra Tor Halt was added on 2 March 1936, initially for workmen at an adjacent quarry but after the latter's closure, was retained for the walkers. Dick visited the halt on 5 July 1955 during the last summer of the line's existence for it closed in March of the following year. The tor itself is on the right of the photo as a mixed train from Princetown approaches the halt, most likely hauled by a 4500 Small Prairie which by this time had taken over from the earlier 4400 class. The basic facilities of the halt are apparent, and by the side of the shelter can be seen the famous notice which read:

GREAT WESTERN RAILWAY COMPANY.
NOTICE
IN THE INTERESTS OF GAME
PRESERVATION AND FOR THEIR
PROTECTION AGAINST SNAKES ETC.
DOGS SHOULD BE KEPT ON A LEAD

BY ORDER.

The snakes were adders, and it is also clear that other life in the form of cattle also roamed freely over this part of the moor.

RCRC5979 On 7 July 1961, No 5541 is seen on a Tavistock-bound goods at Horrabridge. This locomotive is one of the 4575 class Small Prairies which were developments of the 4500 class locomotives, with larger side tanks to increase the water capacity and could easily be recognised by the sloping tops at the front of the tanks. This is yet another preserved member of the class, being found today on the Dean Forest Railway.

RCRC6017 A scenic view taken on 23 June 1962 from the A386 road showing the line south of Whitchurch Down Platform before it curves into Grenofen Tunnel off the photo to the right. 4575 class No 5544 heads a Launceston to Plymouth service. In the distance, a landscape of fields, hedges, and woods gives way to the heights of Dartmoor. On the left the 196 metre high radio and TV mast on North Hessary Tor above Princetown can be identified. In the centre is South Hessary and Cramber Tors, while on the right the mast is part of the infrastructure of RAF Sharpitor, which was a master transmitting station established in 1942 as part of a system enabling aircraft to fix their positions.

RCRC5982 On 2 May 1961, 5541 is seen again on a Launceston to Plymouth train at Tavistock. Being the largest intermediate settlement and station on the line, the station had a wide range of facilities. A goods shed and yard was behind the down platform on the left, while cattle wagons at a loading dock on the up side can be glimpsed on the right. There were private sidings serving a gasworks and sawmills, and engine facilities including a turntable. The passenger station was noted for its overall roof, beneath which in the down platform can be seen an auto train. These operated some of the services from Plymouth which terminated here. The station received the suffix 'South' in 1949 to distinguish it from the town's other station on the Southern main line to Plymouth, which became 'North' at the same time.

5541

RCRC5991 On 7 July 1961, 5541 leaves Mary Tavy & Blackdown towards Tavistock with an up goods. The London & South Western Railway reached Lydford (then spelt Lidford) from Okehampton in 1874 and opened a connection to the Tavistock and Launceston line there two years later. By exercising running powers over the newly mixed gauge route, the LSWR first reached Plymouth. The Launceston branch was still then owned by the South Devon Railway, although worked by the GWR, and did not fully become GW property until 1878 with the absorption of the South Devon. Because of the difficulties of working the extra traffic over the single track of the branch, it is perhaps surprising that it took until 1890 for the LSWR to open its own route to Plymouth by means of the Plymouth, Devonport & South Western Junction Railway Company. This new line started at Lydford and for the first few miles to Tavistock closely paralleled the Launceston line. Here, just south of Mary Tavy & Blackdown, the double track of what later was the Southern main line was on the east side, and is seen emerging from the cutting which passed behind the GW station. There was no corresponding station here on the Southern route. Remarkably, the characteristic concrete platelayers hut on the Southern line survives today alongside the remains of both route's trackbeds, presumably having found use as a farmer's store.

RCRC5984 A few minutes earlier, 5541 and its train are seen in the platform at Mary Tavy & Blackdown. The station opened with the line in 1865, and until 1941 had public goods facilities with a siding and loading dock which stretched beyond the train below the bank. Until the early 1890s there was also a passing loop and second platform. With the removal of all tracks except for the through single line, the signal box was subsequently closed. Although the service did not end until December 1962, the passenger station looks rather neglected by the date of the photo. The Southern line passes behind the scene in a shallow cutting.

RCRC6025 Custom awaits 5544 on a Launceston train at Liddaton Halt on the section west of Lydford. The halt opened on 4 April 1938 but served only a scattering of farms and other isolated houses. Liddaton Green, after which it was named, remains today only the smallest of hamlets. Beyond the line the land descends to the wooded valley of the River Lyd coming down from Lydford Gorge. On the horizon above the locomotive is High Willhays, at 2038 feet the highest point on Dartmoor, with the only slightly less elevated Yes Tor, after which the Southern West Country class Pacific was named, just beyond it. 23 June 1962.

ABOVE • RCRC5939 Coryton also served many isolated settlements, although the village of Chillaton just over a mile away was somewhat larger. Coryton itself was only a small hamlet, but with a church, and was just as near Liddaton Halt. Coryton station opened with the line in 1865 and had public goods facilities which survived with the passenger service until the end of 1962. For these the siding in the background had a loading dock behind the passenger platform and a cattle dock off to the left. However, as at the date of this photo, 23 June 1962, the rust on the rails indicates that the goods operation was not in regular use. The locomotive is 4500 class No 4555 again, calling with a Launceston to Plymouth train.

OPPOSITE PAGE:

RCRC6006 From its opening to the Launceston terminus in 1865, the Western branch had the town's rail business to itself until 1886 when the LSWR's North Cornwall route from Halwill opened. From 1892 when this began to be extended westwards, Launceston became situated on a through route from North Cornwall to Okehampton and Exeter. Approaching Launceston from the east, the North Cornwall line crossed the Western branch and established its station alongside the GWR one. However, there was no connection between the two lines until 1943 when one was put in as part of a scheme to use the Devon and Cornwall network more effectively for goods traffic in case of enemy action closing any of the routes around Plymouth. In 1952 it was decided to use this connection for passenger trains, close the GW terminus (although it remained open for goods), and divert the Western's Plymouth trains into the Southern station. Consequently, in this view from 23 June 1962, 4575 class small Prairie No 5569 is seen in the Southern's North Cornwall line station at Launceston with the 1015 service to Plymouth. On the right can be seen the goods shed of the GW station which remained open for goods until 1966. This was the last summer of the Launceston line. The end of the passenger service coincided with one of the most notable events in the line's history. On the last day, Saturday 29 December 1962, one of the worst snowstorms ever to occur in the west country prevented the last trains of the day from completing their journeys, and in some cases the locomotives and stock were not recovered until into the following week.

5569

6878

RCRC0581 Returning to the main line, this is a panorama over Laira Junction taken from the coal stage of Laira MPD on 17 July 1960. Coming in from Tavistock Junction alongside the Plym estuary, the main line curves under the bridge carrying Embankment Road and passes off to the left into Plymouth. On the south side, goods and relief lines are provided through the junction. Between the road bridge and Laira Junction Signal Box can be seen the gated flat crossing of the original 4'6" gauge Plymouth & Dartmoor Railway which passes behind the carriages in the sidings on the right and parallels Embankment Road to its original termini at Sutton Harbour and Laira Wharf. After the upper section of the P&D closed around 1900, this section was used by the Lee Moor Tramway, the last traffic of which was sand from Marsh Mills to the concrete works near Friary Junction. This ended, horse-drawn to the last (hence the boarded crossing) on 29 August 1960, some six weeks after the date of the photo. By means of an east to south curve by the bridge and an extra rail, the South Devon Railway used the P&D for its first access to Sutton Harbour. However, this was superseded by its own route alongside the P&D in 1878, and this can be seen curving away from the goods lines by the signal box. The foreground is dominated by the entrance to Laira depot. A line of locomotives queue for their turn at the north side of the coaling stage, and in the summer timetable it was not unusual for there being so many that they fouled the through goods tracks as here. The line appears to include a Hall, Grange, Castle, and King, with 6878 LONGFORD GRANGE (82B St Philips Marsh) clearly identifiable on the left.

RCRC0599 On the same day, the various parts of Laira shed are seen before the whole was swept away to be redeveloped as a modern diesel depot. Opened in 1901, the original brick roundhouse can be seen on the right. This had a central turntable and 28 radiating roads. On the left is the four-track 'Long' or 'New Shed' opened in 1931, being one of the 1930s improvements undertaken by the GWR with the assistance of a government loan. At this time of transition, the shed was being used for the diesels in an attempt to create some form of segregation between the incompatible forms of motive power, when it came to servicing and maintenance. Six D6300 type 2 diesel hydraulics are visible, together with a D800 Warship and a D2000 series shunter, later designated as 03. The two identifiable steam locos are Modified Hall No 6964 THORNBRIDGE HALL (84G Shrewsbury) and Hall No 5981 FRENSHAM HALL (82C Swindon).

6873

RCRC0601 The Laira coal stage is seen on 27 August 1961. This was enlarged in 1931 as part of the improvements which included the 'New Shed'. Locomotives could be coaled from either side of the line to the hoist on which the 1361 class saddle tank No 1363 is placing the coal wagons. On the siding in front of the south side of the stage is No 6873 CARADOC GRANGE, at its home shed, and a 2884 class 2-8-0.

5914

OPPOSITE • RCRC4176 A portrait of Hall No 5914 RIPON HALL at Laira shed on Sunday 9 July 1961. The loco was built in July 1931 and withdrawn in January 1964. Earlier a long term Worcester engine, by now it was at 81C Southall, and had probably worked down the previous day on a summer Saturday train when Old Oak Common was short of power. Ripon Hall was originally founded in 1898 in the small Yorkshire city as a hostel for theological students. Moving to Oxford in 1919 it became an Anglican college where its new location was more appropriate for inclusion in the GWR Hall naming sequence.

BELOW • RCRC3999 At its home shed of Laira on 30 August 1961 is No 6400, the first of Collett's 6400 class passenger pannier tanks with smaller wheels than the 5400 series. No 6400 was built in February 1932 and for most of its life was a South Wales engine, moving to Laira for its last few years. As with all the class, it was fitted for auto train working, and was one of the last locos to work on the Saltash service. It was withdrawn in April 1964. The loco is standing on the north side of the coaling stage, beyond which passes the main line west of Laira Junction. Behind is the Plymouth suburb of Laira with the church of St Mary the Virgin on the right. Many of the early terraced houses here were built for railway workers when the shed was first opened, with later council housing visible at the top of the hill.

RCRC4253 The date is 4 July 1957 and the principal train connecting Cornwall with London is departing from Plymouth North Road. The headboard on the train engine, Laira's No 6029 KING EDWARD VIII, reads 'Cornish Riviera Limited'. This headboard was introduced with the chocolate and cream coaches in 1956. Surmounting the headboard, with the usual brown letters on cream background, is the arms of the Duchy of Cornwall with the motto 'One and All'. With this headboard the train reverted to its original 1906 name after periods as 'Cornish Riviera Express' or simply 'Cornish Riviera'. In the public timetable, the train was non-stop from Plymouth to Paddington, but in this instance there would be an operational call to detach the pilot for the south Devon banks. Normally this would be at Newton Abbot but in this case it could be Taunton as the pilot, No 4971 STANWAY HALL, was allocated there. Behind the train can be seen a Castle-hauled service in the opposite platform of the island. This is probably the terminating service which preceded the 'Riviera' from Penzance picking up passengers from the intermediate Cornish stations at which the latter did not call. The rebuilding of the old wooden Plymouth North Road station is under way. This was a protracted affair as it had been started in 1938, stopped by World War Two, and not resumed until 1956. It was completed in 1962. A through station on the GWR main line, it was opened on 28 March 1877 to avoid through trains to Cornwall having to reverse at Millbay. However, the station itself was jointly owned between the Great Western and London & South Western, later Southern, Railways from opening to 1947. LSW/Southern trains served the station by using the GW main line, first from the east to reach Devonport and later from the west to access their own station at Plymouth Friary. After the closure of the latter on 15 September 1958, they terminated at North Road although this had officially lost its suffix on the same date. North Road East Box seen on the right was a replacement dating from 1939, and was closed in 1960 with the opening of the Plymouth Panel Box. Beyond the station and the junction to Millbay can be seen Cornwall Loop Viaduct on the main line westward to Devonport and Cornwall.

MODERNISATION
RECONSTRUCTION
CORNISH RIVIERA EXPRESS

OPPOSITE • RCRC4254 Taken from the same Sutherland Road bridge but looking in the opposite direction, the train is seen entering the cutting through to Mutley Tunnel. Despite the 'Cornish Riviera Limited' headboard, the coaches carry 'Cornish Riviera Express' roofboards. A matching 'Express' headboard was not introduced until 1958, specifically for use with the new diesel hydraulic locomotives whose first duties were on this particular service. Nevertheless, these photos indicate that the distinction between 'Limited' and 'Express' at various times was not always rigid. The sign on the left informs incoming passengers of the North Road station reconstruction. The building on Apsley Road behind, dating from 1901, was the Plymouth Royal Eye Infirmary. These facilities were moved to Derriford Hospital in 2013, and today the Grade 2 listed building has been transformed into a complex of luxury flats.

BELOW • RCRC3811 On an unrecorded date, two of the 1361 class of Churchward 0-6-0 saddle tanks, Nos 1363 and 1361, are seen at work in Plymouth Millbay Docks, the type of duties for which they were especially created. The Plymouth Great Western Dock was first developed by Brunel in the 1840s. A rail connection to the South Devon Railway near its Millbay terminus (which had opened in 1849) was first used in 1850. The docks passed into railway ownership in 1874 and this lasted until after nationalisation in 1948. The docks developed over the years with a wide range of commodities being imported and exported in larger ships than could use the nearby Sutton Harbour. Between the 1870s and World War Two, Millbay Docks were also famous for the USA Ocean Liner traffic when passengers, leaving the liners by tender, could shorten the journey to London by taking the special services to Paddington. Rail traffic to the docks ended in 1971, and today they are the centre of a regeneration area with a marina and Brittany Ferries being the main commercial activities.

RCRC3374 As with the Teignmouth sea wall, Dick also contributed photographs from another very familiar railway location in Devon and Cornwall, the Royal Albert Bridge. At the Devon side on 17 May 1960, a pannier tank is seen leaving the single track across the bridge in charge of a Saltash to Plymouth auto train. This was the last of the Plymouth auto train services, its survival due to the lack of a direct bus service in the absence of a road connection across the Tamar. In fact, from four weeks later, on 13 June, most of this service was operated by DMUs. However, whether steam or diesel operated, the total demise of this intensive service is foretold as the towers of the Tamar Road Bridge are seen under construction. The box in the foreground controlling the end of the single line section was named Royal Albert Bridge Signal Box.

5069

LEFT • RCRC7402 On 28 August 1961, the curve of the Royal Albert Bridge approach spans are seen on the Cornish side at Saltash. The train leaving the single track section and entering the station is again the 0530 from Paddington via Bristol but by now this was only advertised as having through coaches to Plymouth. In fact they then continued after a 45 minute wait at Plymouth as a local service to Penzance, indicated by the headlamp position on the locomotive. As was often the case, chocolate and cream coaches have here strayed into not only a non-titled train but a particularly slow one. Most appropriately for the location, power is provided by Castle class No 5069 Isambard Kingdom Brunel, a Laira engine at this time. It was built in June 1938 and received a double chimney in November 1958. Withdrawal came some six months after this photo in February 1962. By the date of this photograph the Tamar Road Bridge is in the final stages of preparation, and it was first used two months later on 24 October. It was formally opened on 26 April 1962 by Queen Elizabeth, The Queen Mother.

BELOW • RCRC3993 In the down platform at Saltash station, 6400 class pannier No 6400 is seen taking on water before working its train. The service is the 1325 through to Tavistock South, one of the few which were still steam auto trains after 13 June 1960. However, being taken on the same day as the previous photo, 28 August 1961, the intensive suburban service to Saltash, whether steam or diesel, had only a short time left to operate, as after the opening of the road bridge, the station was only served by a much reduced number of stopping trains proceeding further into Cornwall.

RCRC4332 On 18 July 1955, an 82B St Philips Marsh Grange, No 6867 PETERSTON GRANGE, departs from Liskeard and crosses Liskeard Viaduct with an up train. The main line is seen in the distance curving its way towards Menheniot, while between the trees on the right can be seen the Looe branch descending steeply to Coombe Junction. Of the thirty four original timber viaducts on the Cornwall main line between Plymouth and Truro, Liskeard was the second highest of these at a maximum of 150 feet. As with the majority, it was originally built with masonry piers up to 35 feet below rail level from which the timber superstructures fanned out to support the deck. These were replaced here in 1894 with brick piers being added on top of the original stone as clearly seen in the photo. The viaduct was 240 yards long and the fact that it was required to cross not a major valley but a minor stream descending to the Looe River is an indication of the country through which the Cornwall Railway was built and why so many viaducts were necessary.

RCRC4295 No 6828 TRELLECH GRANGE (83F Truro) comes off Liskeard Viaduct and enters Liskeard station with a down goods on 9 July 1961. The signal box at this east end of the down platform replaced the original box on the up platform in 1915. Behind the cream wooden fence beyond the parachute water tank on the left is the footpath leading from the up main line platform to the Looe branch platform. This is at right angles to the main line and the roof of the building on the platform can be seen on the left. The locomotive is crossing the connection to the Looe branch used by goods trains which curves around through sidings and a small goods yard with a goods shed which has the larger roof seen on the left.

RCRC5929 4500 class Small Prairie No 4552 (83E St Blazey) is running round its Looe branch train by Liskeard Branch Signal Box on 17 July 1960. The track in front of the box enters the single branch platform off to the left, and the locomotive is on the run-round loop crossing the points to a siding. On the right are the lines forming the connection from the main line which join the branch just beyond the second signal. Beyond the signals on the right, the branch begins its 180° curve and descends beneath Liskeard Viaduct to Coombe Junction. Liskeard Branch Signal Box was closed in 1964 and replaced by a ground frame where the connection from the main line joins the branch.

RCRC0582 A panorama of Moorswater taken from the main line Moorswater Viaduct on 9 July 1961. This was the site of the head of the Liskeard & Looe Union Canal opened in 1828. The first railway here was the Liskeard & Caradon of 1846 which brought down copper and tin ore and granite for onward transit by the canal for shipment from Looe. In 1860, the Liskeard & Looe Railway replaced the canal and the two railways established their headquarters at Moorswater as the centre of an independent little empire. There was no connection with the outside world until the connecting link was built from Coombe Junction to Liskeard in 1901 and the system was absorbed into the GWR from 1909. The Caradon line closed in 1916. Its remains are the siding passing to the right of the engine shed where the wagons are standing and beyond which its course can be followed below the trees. To the left of the engine shed is a building that was once a carriage and wagon works, an indication of the self-sufficiency that was possible here. In the foreground are goods sidings forming what remains of the public goods station. This would remain open, largely for coal, until December 1963. A full coal wagon can be seen standing on the siding which once extended into the wagon works. A passenger platform, located out of sight below the photo, was opened in 1879 for the service to Looe but closed in 1901 when the connection to the main line was opened and all branch passenger trains to Looe ran from Liskeard, reversing at Coombe Junction. Private goods sidings were important here, serving at various times lime kilns, a granite works, and Cornwall County Council. However, the most important were the china clay dries opened in 1904 and receiving clay slurry piped from Parsons Park on Bodmin Moor. These are off the photo to the left served by a siding unseen below the curve of trees in the left foreground. Originally the clay was shipped from Looe but later all was exported through Fowey. This traffic ended in 1997 but intermittently since then the site has been used as a cement distribution terminal.

RCRC5921 A close up of the engine shed at Moorswater on 18 July 1960 with 4552 and its goods train standing on the siding forming the remains of the course of the Liskeard & Caradon Railway. Moorswater was a sub-shed of St Blazey, and there were workshop facilities at the back of the two road shed. In later years two Small Prairies normally worked from here to cover Looe branch passenger and goods duties. The shed closed in September 1961 when the Looe branch passenger service went over to DMU operation and D6300 class diesel hydraulics handled the goods.

4552

RCRC6012 Nearly the whole 318 yards length of Moorswater Viaduct is visible in this photograph, taken on 9 July 1961. At a maximum height of 147 feet, this was the third highest of the Cornish viaducts after St Pinnock and Liskeard, carrying the main line over the valley of the Looe River. Moorswater was one of the viaducts that were completely new when the original timber viaduct was replaced, in this case in 1881. As can be seen, some of the masonry piers of the original viaduct were left standing. Because of this, and the fact that it crosses the railway to Moorswater, this viaduct has probably become the most frequently photographed on the line. Taken from the road overbridge just to the north of Coombe Junction Halt, St Blazey's 4575 class Small Prairie No 5539 is seen heading towards Moorswater just beyond the viaduct with an engineer's train. This is yet another locomotive of the type that entered preservation after withdrawal in April 1962. It was one of the last to leave Woodham's scrapyard at Barry, going first to the Llangollen Railway and subsequently undergoing restoration at the Barry Tourist Railway.

ABOVE • RCRC5978 No 5539 is seen shortly before the last photograph, waiting at Coombe Junction before proceeding to Moorswater with its engineer's train. Coombe Junction Halt is seen on the adjoining track. After completion of the link line to Liskeard in 1901, this was the reversing point for passenger trains to and from Looe, although in more recent years fewer trains have called here, the reversal taking place at the physical junction by the former signal box off the photo to the right. The halt was actually opened in 1879 with the passenger service from Moorswater to Looe, although it did not appear in the timetables until 1901. Before the service was diverted to Liskeard it was often preferred by passengers instead of Moorswater, as the walk to connecting trains from the main line station at Liskeard was shorter and easier. On the left can be seen the overbridge from which the previous photo was taken, which retains its original arch, and part of Moorswater Viaduct. The building with a chimney behind the halt is not, as might be imagined, anything to do with Cornish mining, china clay, or quarrying. It was the Duchy Tweed Mills, established in the 1880s for the processing of Cornish wool for blankets, clothing material, and the despatch of scoured wool to Yorkshire.

OPPOSITE PAGE:

TOP • RCRC5911 On 10 July 1955, Small Prairie No 4508 is about to pass the Looe distant signal on the final approach to the terminus with the branch train from Liskeard. What more is there to say about this idyllic Cornish scene beside the Looe River? Except perhaps to note that the original Liskeard & Looe Union Canal only began some three quarters of a mile upstream at Terras Pill and therefore this final section in the river on to Looe could not have been easy to navigate at low tides such as seen here.

BOTTOM • RCRC5935 On 18 July 1960, the single passenger platform at Looe is host to 4552 on the goods train seen earlier at Moorswater. The broad swathe of grass separating the platform from the Looe River gives ample opportunity for reinforcing the name in stone. The line continued beyond the passenger terminus to a small goods yard, engine shed, and carriage shed, and further still to a quayside owned by the Harbour Commissioners. In earlier years the latter was busy with the ores and granite from Caradon and china clay from Moorswater which were brought down by the railway, but after many years of decline was last used in the early 1950s. The public goods service to Looe ended in November 1963, but thankfully the delightful branch and its passenger trains are still with us, although not the station building shown here which has been replaced by a small modern, but not unsympathetic, structure.

ABOVE • RCRC3584 Scenic from one end to the other, nevertheless arguably one of the most beautiful sections of the main line through Cornwall is the six mile descent from Doublebois to Bodmin Road through the well-wooded Glynn Valley of the River Fowey. The railway follows a course half way up the southern side of the valley, giving a much greater panoramic view of the surroundings than from the A38 road following the valley floor. Tributary valleys are today crossed by no less than seven of the line's viaducts, although originally there was even one more which was replaced by a retaining wall and embankment instead of being rebuilt. Seen here is St Pinnock, the second viaduct down from Doublebois, and at 151 feet the highest of all the Cornish viaducts, although only one foot higher than Liskeard. It has a length of 211 yards and the timber superstructure was replaced by masonry on the same piers in 1880. Today, St Pinnock is the most accessible of the viaducts to appraise as it strides across the rear public car park of Trago Mills. Actually beginning at Lostwithiel, the eastbound climb was one of the most taxing of the many adverse gradients on the Cornwall main line and heavy summer Saturday trains often had to be double-headed. In this undated view an unidentified Grange in BR black mixed traffic livery has an easier task on the climb with a Plymouth-bound local train.

OPPOSITE • RCRC3354 Seen here on 15 June 1960, Bodmin was often regarded as one of the GWR's finest branch termini and an ideal one for modelling. Opening in 1887, it became Bodmin General in 1949 when the Southern station across town became North. On the left is the single passenger platform with the signal box on the ramp. The station buildings also extended at right angles to the line as it was originally intended that a second platform would be built on the run-round loop. That has now been achieved in preservation. On the right is the goods shed and yard which today are the workshops of the Bodmin Railway, and on the nearer left the single road engine shed and parachute water tank. Add the trees and the rhododendron bushes on the banksides and the atmosphere is complete. Added interest is generated by the fact that two routes entered the terminus. The line in the centre is the original 1887 route from Bodmin Road. Diverging to the left is the 1888 extension to join the London & South Western's Bodmin & Wadebridge line at Boscarne Junction. This was the latter's first connection to the outside world, and the only one until the North Cornwall line reached Wadebridge in 1895. It enabled GW passenger trains from Bodmin Road to run through to Wadebridge and Padstow, as well as interchange goods services which included china clay from Wenford Bridge. Because of the physical layout, but adding to the operational interest, all this through traffic necessarily had to reverse in the terminus. The passenger services ended in January 1967 and general public goods three months later, but clay traffic from Wenford Bridge kept the railway active here until September 1983. Thankfully, the whole system from Bodmin Road (now Parkway) to Dunmere Junction now operates as the preserved Bodmin Railway with its headquarters here at General.

LOSTWITHIEL
CHANGE FOR FOWEY

OPPOSITE • RCRC3865 8750 class 0-6-0PT No 9655 (83E St Blazey) crosses from the down to the up main line at Lostwithiel on 23 September 1960 with a train of empty china clay wagons from Fowey, probably bound for Wenford Bridge and Marsh Mills. The bay platform for the branch train to Fowey was on the far side of the down platform to the left beyond which were sidings normally used by clay wagons bound to and from Fowey Harbour. The Fowey branch itself verged left from the main line beyond the bridge over the River Fowey which can be seen to the left of the Toad brake van on the rear of the train. Beyond the wooden goods shed behind the clay wagons and through the cab of the pannier tank can be seen the ends of the buildings which were part of the original 1859 Carriage and Wagon Works of the Cornwall Railway. The buildings were closed for this function by 1920, although remained in railway use until the 1950s. Today they are listed buildings and have been converted into modern apartments.

RIGHT TOP • RCRC3502 The station at Fowey is seen on 22 July 1960. Modernisation is evident as Warship No D816 ECLIPSE brings a train of empty clay wagons around the bend from the harbour bound for St Blazey. The locomotive was only a few months old, being introduced in the February. The Fowey branch passenger train has been recessed into the bay platform normally used as a loading dock. The Fowey service was unique among the Cornish branches at this time as still using an auto train and 1400 class 0-4-2 tank for most of the trains. Seen here is No 1419 of St Blazey, the regular loco from the early 1950s. On the right is the goods shed and yard with some full china clay wagons awaiting movement to the harbour.

RIGHT BOTTOM • RCRC6018 4575 class Small Prairie No 5572 is portrayed in the platform at Fowey on 23 September 1960. The Laira locomotive was working the branch train to Lostwithiel. These Prairies or occasionally Pannier tanks were called in to work the service when the 1400 tank was under repair. Obviously unknowingly at the time, Dick seems to have a predilection for photographing Prairies that would later enter preservation as this is yet another example, being found today at Didcot Railway Centre. Ahead of the train continues the direct line to St Blazey. This was used by a public passenger service until 1929, after which it was principally used by china clay trains to Fowey Harbour and the return of empties to St Blazey.

ABOVE • RCRC8040 On Saturday 9 April 1960, Dick enjoyed a footplate trip on Newton Abbot's No 4083 ABBOTSBURY CASTLE on the 1330 Penzance to York broccoli special. It is seen here approaching Par, descending the short but sharp 1 in 57 gradient from Par Viaduct over the line to the harbour and the direct route from St Blazey to Fowey. The train is signalled into the up main line platform with the signal box at the south end. The far side of the up island is used by the Newquay branch trains, and to the left is the wooden goods shed. Curving away to the left is the start of the Newquay branch as the loop around to St Blazey. This was opened by the Cornwall Minerals Railway in 1879 to give a connection from its system to the Cornwall Railway main line in Par station. However as the Minerals railway was standard gauge and the main line broad, there was no through working here until the conversion of May 1892. The two sidings being passed on the left were generally used for carriages and vans. ABBOTSBURY CASTLE was built in May 1925 as the first of the second batch of production Castles and was withdrawn in December 1961.

BELOW • RCRC3596 On an unidentified date, a train of empty china clay wagons returning from Fowey to St Blazey approaches Polmear and Par Sands on the descent from Pinnock Tunnel. The locomotive is one of the Churchward 4200 class 2-8-0 tanks. After 1929, at least two of this class were allocated to St Blazey, mainly for working the china clay trains over this steep gable route between St Blazey and Fowey, although they were often found on other work such as main line turns and as train engine or bankers between St Blazey and St Dennis Junction. This section of route was closed in July 1968 and converted into a private road for English China Clay's lorries between Par and Fowey Harbours, leaving the Lostwithiel line the only rail access to Fowey.

BELOW • RCRC0572 On 20 July 1960, Warship No D816 ECLIPSE waits in St Blazey yard before departing with a loaded train of china clay for export through Fowey Harbour. The building behind the train is the wagon works which maintained and repaired the fleet of china clay wagons serving the area from the days of the Cornwall Minerals Railway, through to today. Off the photo to the left is St Blazey MPD. Although modernisation was underway, the shed did not close to steam until April 1962. Standing on the siding from the turntable are a Manor (believed to be 7812 ERLESTOKE MANOR), a County, and two Small Prairies. St Blazey's allocation was mainly tank engines for work on the china clay lines, but also included 4300 class 2-6-0s and limited numbers of the 4-6-0s over the years. The largest of the latter were two Counties, 1002 COUNTY OF BERKS and 1006 COUNTY OF CORNWALL which were shedded here at the time of the photograph. Accordingly, it is likely that one of these two is the loco illustrated. However, other 4-6-0s were regularly serviced here after working incoming trains, and all could be seen on summer Saturdays undertaking piloting and banking work on the heavy trains to Newquay. To the right of the sidings can be seen the bracket signals for entering and exiting the loop around to Par station, while the Newquay branch continues beyond the wagons past the remains of the former passenger station.

OPPOSITE • RCRC4199 When the Cornwall Minerals Railway Company acquired the Par and Newquay Treffry tramways in 1873 with the intention of rebuilding and connecting them, the largest part of the work was to replace the original route of the Par Railway via the rope worked Carmears Incline in the Luxulyan Valley by a new route nearer the valley floor. However, this was only achieved by a two and a half mile constant climb up the valley, most of which was at 1 in 37. This may not have been too serious for the ordinary Newquay branch train, but it was a different matter in the latter days of steam for the heavy summer through trains to the Atlantic coast resort. On 5 July 1955, the 0930 Paddington–Newquay is seen making the climb double headed by 5972 OLTON HALL and 6397, banked by 5519. The Hall was a Penzance engine but both 6397, a 4300 class 2-6-0, and 5519, a 4575 Small Prairie, were St Blazey locos employed here on the regular summer duties for this shed. OLTON HALL is preserved and first steamed at Carnforth in 1998. However, it subsequently became famous and well known to thousands of youngsters (and many much older) not otherwise interested in railways as the crimson-liveried HOGWARTS CASTLE used in the Harry Potter films. To more rail-minded people it became 'The Hall that thinks it's a Castle'.

RCRC5989 Having climbed the side of the Luxulyan Valley by means of the Carmears Incline, the Par Railway reached Luxulyan by crossing the valley on the Treffry Viaduct. Today the viaduct provides a magnificent vantage point of the replacement route at the upper end of the valley. Here, photographed from the viaduct on 20 July 1960, 4575 class Small Prairie No 5541 descends from Luxulyan with the branch train from Newquay to Par. As previously indicated, this is one of the later preserved members of the class.

RCRC3875 St Dennis Junction was a significant location on the Newquay branch and a key point in the operation of the lines serving the Hensbarrow china clay district. The junction is seen from the old A30 road bridge in July 1955. What appears to be a double track line straight ahead is in fact two single track routes. The left track is the line to Drinnick Mill and Burngullow on the original formation of Treffry's Newquay Railway serving many china clay works along the way. The right track, descending behind the bushes on the right, is the Retew branch to Meledor Mill serving china clay works by the River Fal. Neither of these lines have ever had a passenger service. The four sets of wagons, from left to right, are standing on the tip sidings, the up and down loops off the Drinnick Mill route, and the loop off the Retew branch. On the left, the Cornwall Minerals line of 1874 from Bugle, connecting the Par and Newquay Railways and completing the through line to Newquay from Fowey, and later Par, curves in after crossing Goss Moor. This section on the moor was doubled in 1921 to improve operations on the Newquay branch especially during the summer service. The line on to Newquay reverts to single track behind the photographer immediately after the bridge. The Par to Newquay branch train taking the curve and signalled through the junction is headed by 8750 class 0-6-0PT No 9673 of St Blazey. In the background, the landscape of Hensbarrow Downs is punctuated by the waste tips of the china clay industry.

ABOVE • RCRC3577 In July 1955, a down train hauled by a Hall 4-6-0 is seen on the 1 in 61 climb past Par Harbour. The identity of the three-coach train is unknown but with roofboards on the carriages it could be a portion of a long distance train rather than a local service. The line of vans beyond the train are standing on sidings beyond Par Harbour Junction, whose signal box can be seen, from which there was direct access to the harbour from the main line, the other access being from St Blazey at the far end. Par was one of the two ports which exported china clay, but unlike Fowey could only accommodate smaller vessels which accordingly were mostly involved in the UK coastal and near-Europe trade. The harbour had its own complex of clay dries with slurry piped down from the pits on Hensbarrow. The increase of this operation over the years led to a corresponding decrease of rail transit to the harbour. On the left can be seen one of the dries of the Par Moor complex on the opposite side of the main line to the harbour. These were accessed by a line using a low bridge under the main line which necessitated the use of the cut-down locomotives in the harbour.

BELOW • RCRC3505 Grampound Road was one of the less photographed stations on the Cornwall main line. It is seen here on an unknown date with the signal box mid-way on the up platform and the goods shed and yard beyond. The station opened with the line in 1859 and in GWR tradition the 'Road' suffix indicated that it was some way from the settlement it purported to serve. In this case Grampound was about two miles away, but as in some other cases, a later settlement grew up around the station and taking its name from it. It also served several other scattered locations of which Ladock was actually closer than it was to Probus & Ladock Platform, which was opened in 1908 some two and a half miles further west. Grampound Road closed to goods on 1 June 1964 and to passengers on the fifth of the following October, the same date as several other stations on the main line, leaving no passenger station between St Austell and Truro.

4593

RCRC5969 On 8 April 1960 4575 class Small Prairie No 4593 is seen leaving Truro with a branch train for Falmouth. It is actually departing from the down main line platform rather than the bay normally used for the branch service. The points leading into the latter are beneath the engine and first coach and above the grass bank can be seen the bay platform signal at danger with the main line signal at clear. Beyond is the up island platform and a range of sorting sidings, the goods yard being out of sight on the down side at the far end of the station. On the left behind the West Signal Box is the entrance to 83F Truro motive power depot.

OPPOSITE • RCRC8045 Penwithers Junction, just west of Truro, was one of those locations on the UK rail system with a somewhat unusual history that make them such delightful locations to study. It is seen here on 20 July 1960 looking along the Falmouth branch, past the main line curving in from Penzance and through Higher Town Tunnel and the cutting to Truro station beyond.

The first railway here was the standard gauge West Cornwall which opened for passengers from a temporary terminus at Truro Road, off the photo to the left to Redruth and Penzance in 1852. Goods services began in 1855 at the same time as which an extension for goods and passengers from Truro Newham by the river was opened to Truro Road, cutting across the area in this photograph from right to left. Truro Road was closed at the same time. The broad gauge Cornwall Railway had been authorised throughout from Plymouth to Falmouth in 1846 and was opened as far as Truro in 1859.

At the same time the connection between the two railways at Truro was made with a curve of the West Cornwall's from the Newham line on the left and the opening of the Cornwall Railway through the tunnel which was mixed gauge track. (The whole of the West Cornwall's route was later mixed gauge.) The first physical junction at Penwithers was therefore off the photo to the left.

The Cornwall's opening on to Falmouth came in 1863 which involved the second physical junction at Penwithers, the one seen here, and a flat crossing over the Newham line in the centre of this photograph. The latter was closed to passengers at the same time and all passenger trains used the Cornwall's Truro station.

In 1894 the GWR, now the owner of the wholly standard gauge system, eliminated the flat crossing by closing the access to the Newham goods branch from the first Penwithers Junction on the Penzance line and replacing it with a connection from the Falmouth line at Newham Branch Junction. In the photo Dick is on a goods van behind 4575 class No 5552 leaving the Falmouth line at this point to travel down to Newham. The original ownership explains why the Falmouth branch is straight on at this location and the main line to Penzance curves away.

Of the four compass points at which route formations have joined here, only one, that to the east, was never a physical junction as the 1894 change meant that one route opened as the other closed at this point. What seems inexplicable is why the new 1894 connection faced Falmouth when there seems ample room to have made a more logical direct connection facing the tunnel. The Newham branch closed in 1971 when the signal box here, just off the photo to the left, also closed, the remaining Penwithers Junction (where the Falmouth branch left the main line to Penzance) being controlled from Truro.

RIGHT • RCRC8046 Dick is approaching the goods terminus at Truro Newham behind 5552 later on the same journey. The sidings in the goods yard are seen straight ahead with the backdrop of Truro Cathedral. A back shunt siding passing behind the pile of coal served a jetty on the Truro River to the right of the locomotive. Remarkably, Dick has here photographed another Small Prairie which came to be preserved. A Truro engine for the last few years of its BR life, 5552 has now been fully restored and is working, most appropriately, on the Bodmin Railway. It is also of sentimental interest to the author as it was the first western engine he underlined in his combined volume as a ten-year old, seen in 1958 from a car on the A30, clanking across Goss Moor on a Newquay branch train.

4564

RCRC5927 At Gwinear Road on 9 April 1960, Truro's 4500 class Small Prairie No 4564 marshals a train of 'Oxfits' loaded with broccoli from the Helston branch before adding them to the 1330 Penzance–York 'Broccolo', as these specials were generally known by the railwaymen. This was the last season when steam would appear regularly on these specials, which for many years had been an important business for the railway as well as for the farmers who could take advantage of west Cornwall's mild winters to market their produce, when it had not yet ripened for sale from other areas.

Organising the wagons, motive power, and destinations of the specials had to be carried out at short notice as prices and demands of the market could change on a daily basis. By this date road transport had captured much of the traffic but at peak times up to seven daily services could still be required, typically two for London, two for South Wales and three for the Midlands or beyond. The latter included individual wagons for forwarding to a wide range of final destinations in the Midlands, northern England, and Scotland.

Most specials started at Ponsandane or Marazion and picked up wagons at other points on the main line, including from the branches. As seen, self-ventilating 'Oxfit' cattle wagons, suitably cleaned, were still widely used on the specials, although covered 'Vanfits' with end ventilation were increasingly used by this time. The broccoli itself was loaded in non-returnable wooden crates and stacked in the wagons.

RCRC8024 Seen from the footplate, Castle No 5053 EARL CAIRNS pauses at Gwinear Road with the 1400 Penzance to Crewe perishables train on 29 April 1961. This loco was built in May 1936 as BISHOP'S CASTLE and renamed in August 1937. It was withdrawn in July 1962. This was another 'Road' station but in this case Gwinear was a very small place, some others were nearer on more direct roads, and no significant settlement developed nearby. However, it assumed greater importance after the Helston branch was opened from here in 1887. The branch passenger service used the far side of the down island platform behind the West Signal Box by the level crossing. The branch is seen diverging ahead while the main line curves to the left. The wagons are standing on a siding which normally served a cattle pen. There was also an East Signal Box which controlled the eastern access to the series of sidings seen in the previous photo between the main line and the branch. These were used to marshal the goods trains from the branch which brought agricultural produce from a wide area of southern Cornwall including most of the Lizard peninsula. The Helston branch closed to passengers in November 1962 and to goods in October 1964, when Gwinear Road itself closed to passengers. It survived until August 1965 for goods.

ABOVE • RCRC5943 The branch from St Erth to St Ives packs more beauty and atmosphere into its four and a quarter miles than most lines ten times as long. The St Ives terminus is seen here above Porthminster Beach, with part of the town near the harbour in the distance. On 24 September 1960, passengers board the branch service to St Erth in the charge of Small Prairie No 4564. There was only one main platform for the passenger trains, with the run-round loop to the right. The track on the left of the platform was mainly used to serve a loading dock on the far side. A further goods siding with a goods shed was behind the photographer, as was the signal box and a single-road engine shed beyond a short viaduct. The station was taxed to capacity on 1950s summer Saturdays when for a period the main portion of the 'Cornish Riviera Express' with restaurant car terminated here.

RIGHT • RCRC8041 Marazion, the last passenger station before Penzance, is seen here on 9 April 1960 as Dick prepares to depart on the footplate of 4083 ABBOTSBURY CASTLE on the 1330 Penzance to York broccoli special previously seen approaching Par. The train, which began its journey at Penzance Goods (Ponsandane), will have taken on extra wagons here from the loading dock behind the passenger platforms to the left. This and the adjoining sidings could be accessed from both ends of the passenger platforms, the connection at the far end just invisible between the platforms and the road overbridge. Marazion was another of the stations which lost its passenger service on 5 October 1964, the public goods service surviving until December 1965. Marazion itself is seen a mile away on the right, a large village on the shore of Mount's Bay and directly opposite St Michael's Mount off the photo to the right.

RCRC0567 83G Penzance MPD was nearer Marazion than Penzance itself, situated between the A30 road and the tracks of the main line running alongside Mount's Bay. It was familiarly called Long Rock after the nearby settlement. The shed opened in 1914, replacing earlier facilities nearer the station. In this view of 9 April 1960, from the water tower by the coaling stage, the four-road shed and repair shop are the main features with the turntable glimpsed on the far left.
The locomotives on view include single examples of Castle, County, and Small Prairie as well as at least four Granges. In 1960, the Granges were the class with the largest numbers allocated here, and there were also Counties, Small Prairies, pannier tanks, and smaller numbers of Castles and Halls. The shed closed to steam in September 1962 and was used by diesels until the whole was swept away to make room for today's Penzance Traction Maintenance Depot.

RCRC0566 Transition at Penzance MPD on 24 September 1960. Grange No 6828 TRELLECH GRANGE and Modified Hall No 6988 SWITHLAND HALL stand outside the running shed with Warship No D833 PANTHER on the line into the repair shop. Both steam locomotives were visitors, the Grange from Truro and the Hall from Bristol Bath Road.

RCRC4289 No 6824 ASHLEY GRANGE, one of Penzance's own, stands ahead of a County on one of the lines between the shed and the turntable at Long Rock. 24 September 1960.

6814

OPPOSITE • RCRC4291 Visiting from Bristol St Philips Marsh, immaculate No 6814 ENBORNE GRANGE is photographed on 29 April 1961 on the sidings by the coal stack at the east end of Long Rock depot.

BELOW • RCRC7372 The last photograph is appropriately at Penzance, the end of the main UK railway system. The only rails further on were those of the Penlee Quarry Railway and mine tramways in the west of the Penwith peninsula. The station is largely in the condition left after the rebuilding of the late 1930s, which included expanding the former cramped passenger facilities and removing most of the goods accommodation to Ponsandane alongside the main line between the passenger station and Long Rock depot. There were two long island platforms and sidings for the remaining goods traffic. The latter consisted mainly of mails, parcels, perishable traffic, and goods to and from the Isles of Scilly via the quay including fuel for the ferries. In this view from 29 April 1961, No 5053 EARL CAIRNS is standing in platform 4 awaiting departure with the 1400 Penzance to Crewe perishables train, seen in a previous photo at Gwinear Road. The first vehicle is an outside framed Siphon G to diagram O11. Platform 4 is the only one wholly outside the overall roof, seen above the locomotive. In front of the train are the sidings to the loading dock and quay. On the far side, Type 2 diesel hydraulics Nos D6307 and D6311 stand in platform 2 on the near side of the other island platform. The line in platform 1 is between the island and the stone retaining wall, above which is the A30 road entering the town and from which generations of trainspotters and others have enjoyed a grandstand view of activities in the station.